Insects and Plants

Developed at
The Lawrence Hall of Science,
University of California, Berkeley
Published and distributed by
Delta Education,
a member of the School Specialty Family

1325263
978-1-60902-055-2
Printing 3 — 7/2013
Quad/Graphics, Leominster, MA

Table of Contents

Animals and Plants in Their Habitats

Look at this grassland **habitat**.
Do you see anything **living** here?
Let's take a closer look.

Look! A grasshopper sits in the green grass.
Grass plants and grasshoppers are living in this
grassland habitat.
Grass and grasshoppers get their **basic needs**
met in the grassland habitat.

Land animals need food, water, air, space, shelter, and a comfortable temperature. Grasshoppers eat grass for food and water. In the grass, grasshoppers have air and shelter. In summer, the grassland provides a warm habitat for **insects**, like grasshoppers, and other animals, like prairie dogs.

Plants need water, air, sunlight, and **nutrients**.

Grass gets water and nutrients from the rich, moist grassland soil.

Their roots take up water and nutrients.

Air flows around the grass.

Grass uses sunlight to make the food it needs to live.

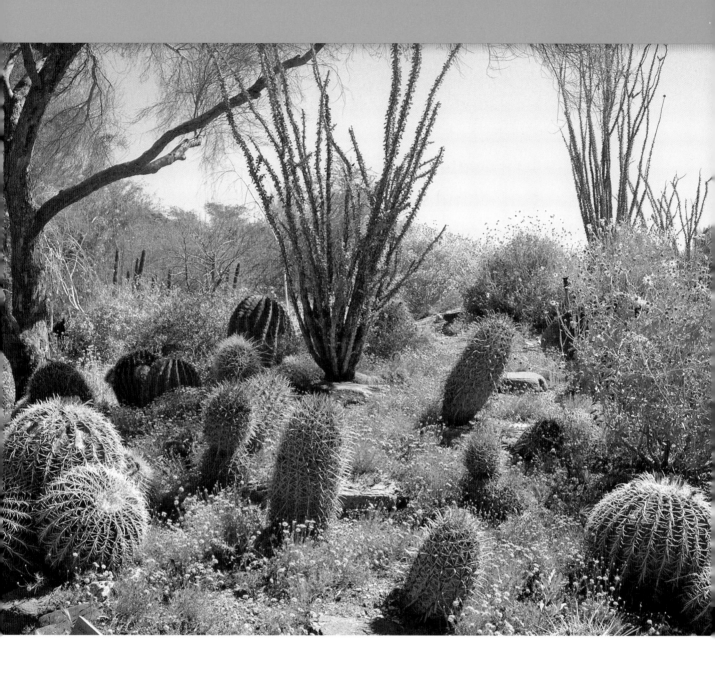

Other parts of the country are different
from the grassland habitat.
This is a desert habitat.
What do you see living here?
Let's take a closer look.

Plants and animals are living in the desert, too.
How do they get the things they need to live?

Desert plants get water and nutrients from the
desert soil.
The plants have large root systems to collect water.
Some desert plants, like cactuses, have thick
spongy stems.
Cactus stems store water for the plant to
use later.

The Sun shines brightly on the desert most of the time.
Air flows easily around the desert plants.
Deserts are very hot during the summer.
But desert plants can live even in very hot temperatures.

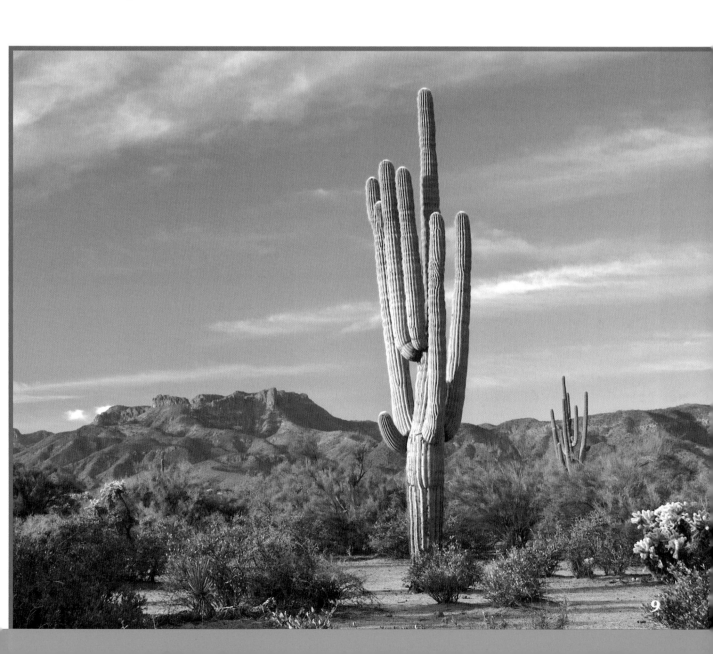

Some desert animals eat other animals for food.
Some desert animals eat **seeds** for food.
Harvester ants gather seeds.
They store the seeds to eat during the year.
Ants get water from the seeds they eat.

Ants dig tunnels and chambers.
Their underground tunnels give ants a safe space
to live.
The tunnels provide shelter for the ants so they can
survive the hot desert summer.

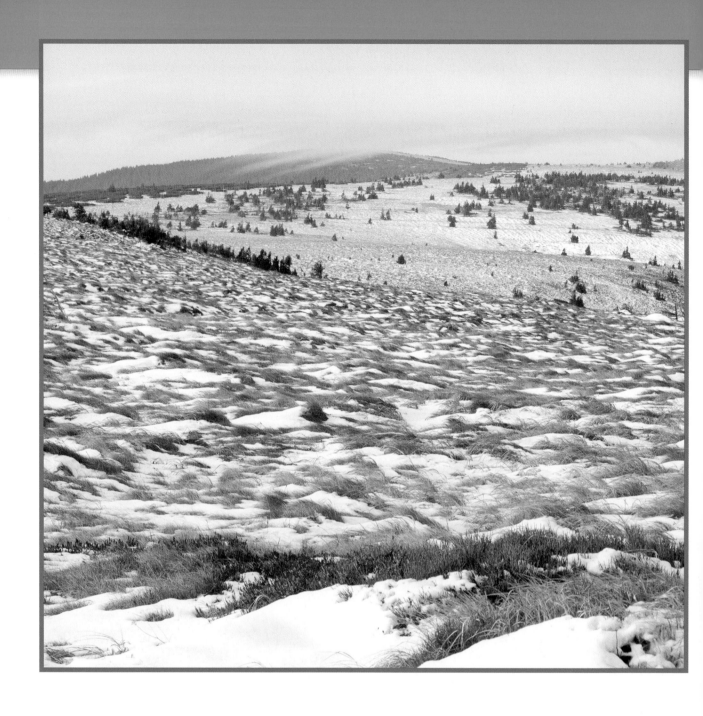

This is a tundra habitat in winter.
Some parts of Alaska have this kind
of habitat.
What do you see living here?
Let's take a closer look.

Tundra plants survive the winter in a resting stage.

When the Sun warms the land, the plants begin to grow again.

The plants take up water and nutrients from melting ice and snow.

They use sunlight to make their food.

Tundra plants can survive cold temperatures.

Caribou roam the open space of the tundra.
They eat the short plants.
They drink from pools of melted snow.
Caribou have thick fur to protect them from the cold temperature.
But the fur is not thick enough to protect them from mosquitoes.

Mosquitoes burrow down in tundra plants
in the fall.
They rest there during the cold winter.
When the snow melts in spring, the mosquitoes
look for food and mates.
The female mosquitoes lay **eggs** on pools
of water.

After the eggs hatch, mosquito **larvae** eat tiny bits of food in the water.

The mosquito larvae grow quickly for a few weeks. Then they swim to the surface of the water and break out of their old skin.

Now they are **adult** mosquitoes and fly to find food.

When you look closely at a habitat, you will find animals and plants living there.
Living things **thrive** when they have what they need to live.

Flowers and Seeds

These are wild brassica plants.
Each plant grows a lot of **flowers**.
But brassica plants do not grow flowers
to look pretty.
The flowers are an important part of
the plant's **life cycle**.

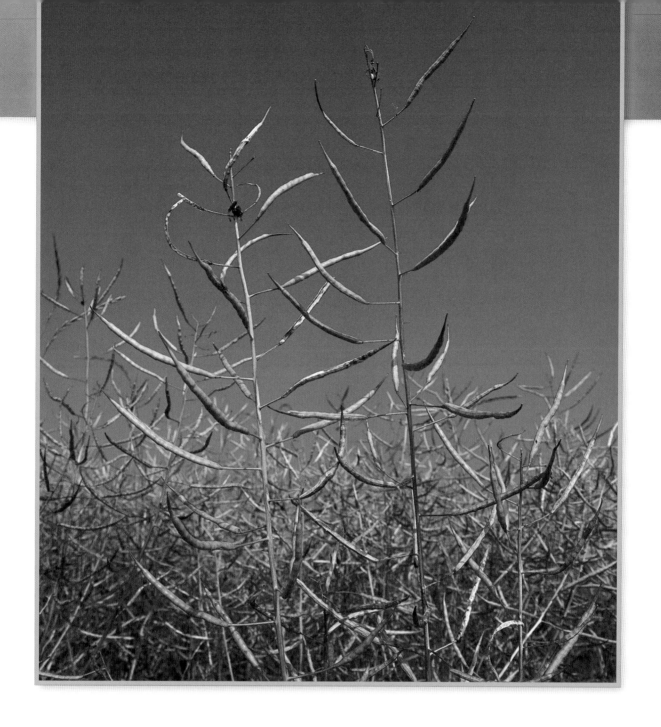

Soon, the flowers fade and dry up.
Something new appears right where each
flower once grew.
It looks like a little green bean.
It is a seedpod.

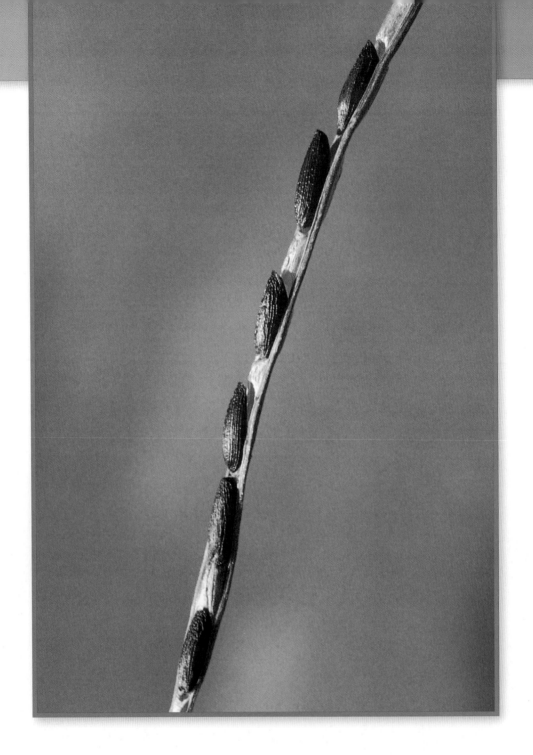

Weeks later, the seedpods are big and dry.
There are about six seeds inside each seedpod.
What do you think will happen if someone
plants the new seeds?

Brassica plants are not the only plants
that make seeds.
Cherry trees make seeds.
Where are they found?

There is one seed inside each cherry.
And where does the cherry grow?
It grows right where the cherry flower was.

Plants grow flowers.
The flowers grow into **fruit**.
Fruit have seeds inside.
When the seeds grow into new plants, it is
called **reproduction**.

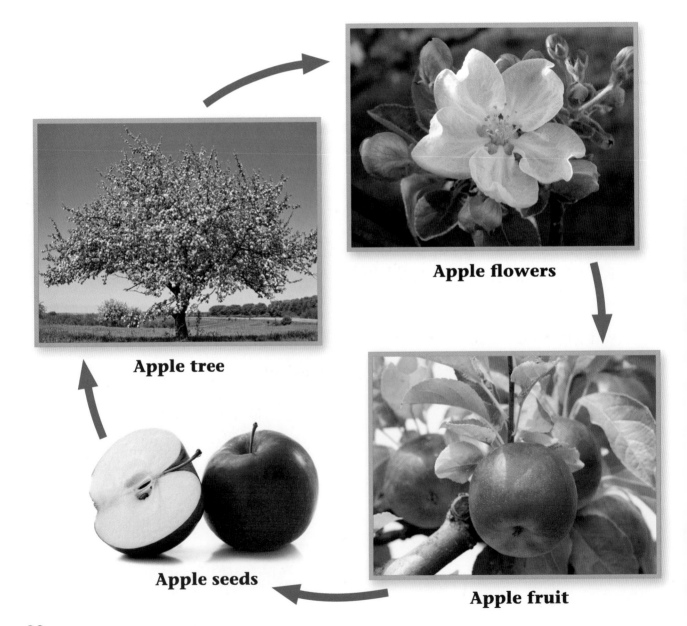

Apple tree

Apple flowers

Apple seeds

Apple fruit

Have you ever seen tomato flowers?
Tomato flowers grow into tomatoes.
Tomatoes are fruit.
They have seeds.

Can you see the strawberry flowers?
Strawberry flowers grow into fruit.
Strawberries have seeds, too.

New plants grow from seeds.
Seeds are found in fruit.
Fruit grow out of flowers.
Flowers and fruit are important in the life
cycles of plants.

Thinking about Flowers and Seeds

1. Name one plant, and tell about its flowers.

2. Where are the seeds on a brassica plant?

3. Name two fruits you like to eat.

4. Name the parts of a plant that are important for its life cycle.

So Many Kinds, So Many Places

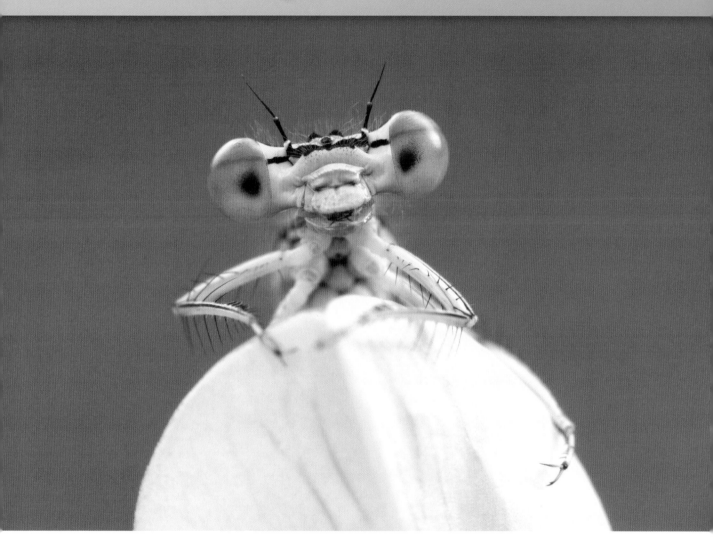

This amazing animal is an insect.

Flies, ants, and crickets are all insects, too.

There are so many kinds of insects.

Insects are everywhere!

Can you name some others?

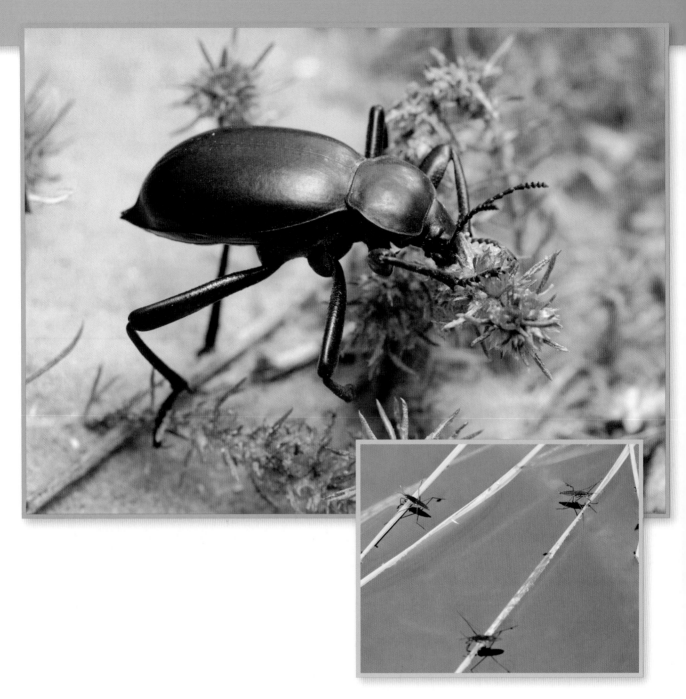

No matter where you are,
an insect is probably near you.
Insects are in the air and in the water.
Some creep in the Arctic snow.
Others scamper around in the desert.

These ladybugs have gathered on a tree trunk.
Some insects live on the tops of mountains.
Other insects live in the rain forest.
Insects are everywhere!

Insects might seem like pests to you.
Some insects eat clothes, buildings, and crops.
But insects are very important.
Many different animals need them for food.

Insects are important for people, too.
Bees visit flowers, and that helps fruit grow.
They also make sweet honey.

People use thread from the cocoon of the
silkworm to make clothing.

Next time you go outside, look for insects.
They are everywhere!

Variation

Variation means difference.
When we looked at darkling beetles, we saw variation.
They were not all the same.
Do other **organisms** have variation?

People are different.

Some people are short.

Some have brown eyes.

Some have black hair.

Some have freckles.

Everyone is different when you look closely enough.

These black bears have the same shape and size.
But like other animals, they have variation too.
What kind of variation do you see?

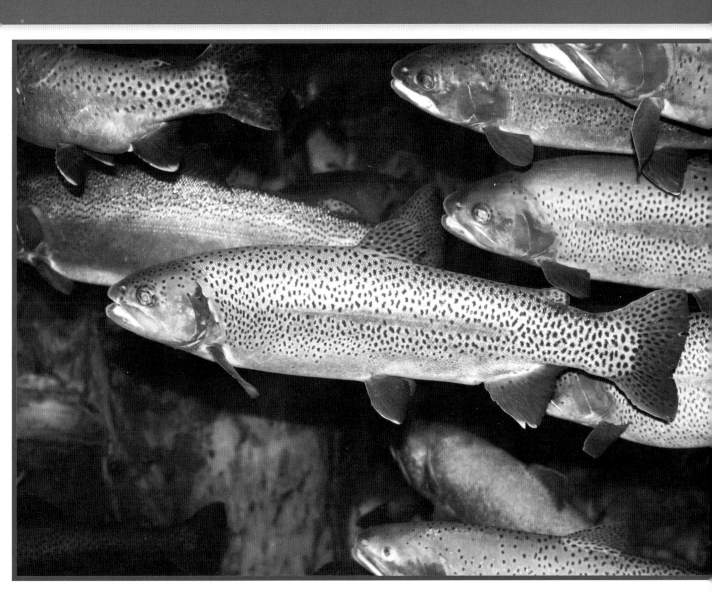

Trout have color variation, too.

Some of them are silver.

Others are brightly colored.

Some have a lot of spots.

Others have only a few spots.

Trout have both color and pattern variation.

These shells are all from the same kind of scallop.
What kinds of variation can you see?

Here is a garden with marigolds.
Do you see any variation?

Thinking about Variation

1. Name five variations you can observe about people.

2. Tell about variations in trout.

3. Think of another animal, and tell about its variations.

Insect Shapes and Colors

Insects are different shapes and colors.
The shape or color can help insects hide.
An enemy might not see an insect that looks
like its habitat.
A hungry bird or lizard might think this
insect is a leaf.

This praying mantis hides in the leaves, waiting to catch an insect to eat.

The walking sticks on this twig are very hard to see. Can you find them?

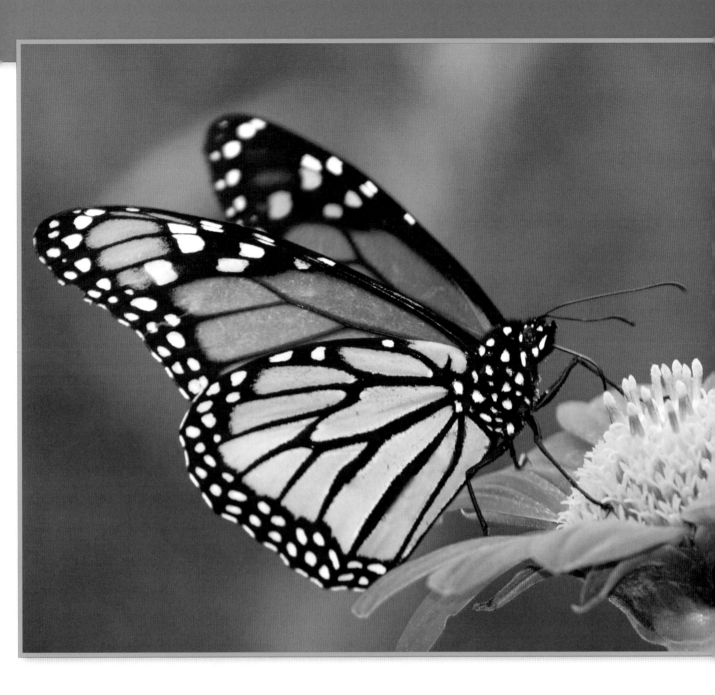

Look at the bright colors and design on
this butterfly.
Do you think it is hiding from its enemies?

Some insects are very easy to see.
They are very colorful.
They might have special markings.

Brightly colored insects often taste bad.
They make other animals sick.
Animals learn to stay away!

The spots on this beetle look like huge eyes.
A hungry animal might think the beetle is
a much bigger insect.
The animal might be scared away.

Insect Life Cycles

Insects might look different at each stage of their lives.

Most insects go through four stages.

The stages are egg, larva, **pupa**, and adult.

The eggs of this insect were laid inside cells.

After a few days, a larva hatches from each egg.
The tiny larva stays curled up inside the cell.
The larva eats food made from pollen and honey.
This food makes the larva grow.

Then, the cell is covered with wax.
Inside the cell, each larva turns into a pupa.

In the pupa stage, the insect goes through
a big change.
Soon, an adult crawls out of each cell.
Do you know what insect this is?

It's a bee!

After a short rest, the bee can go right to work.

Young adult bees work in the hive.

Older bees work outside the hive.

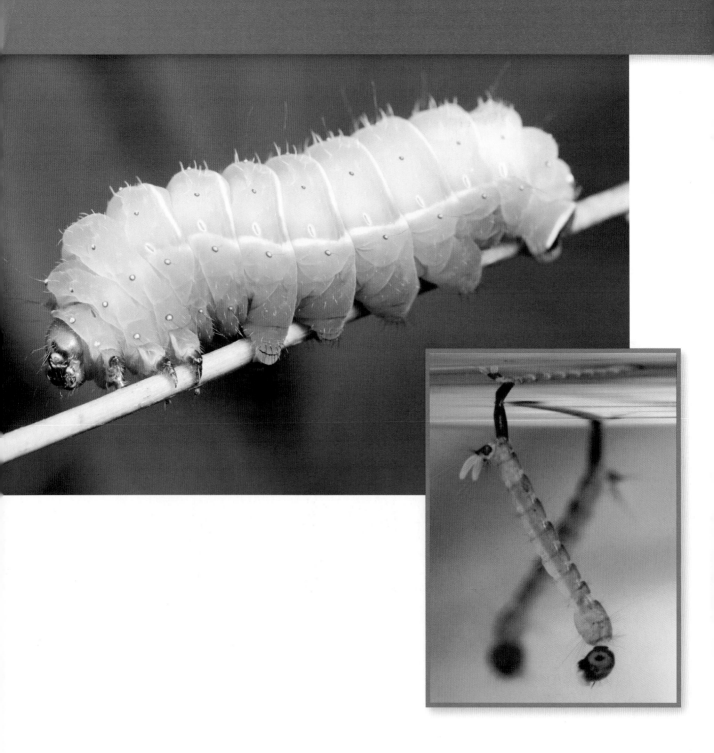

The larvae of different insects do not look the same.
These larvae will become insects you know well.
What will they look like as adults?

Moths and mosquitoes!

Some kinds of insects don't have larvae
or pupae.
When they hatch from eggs, they are
called **nymphs**.
Many nymphs look like their parents,
but smaller.

Milkweed bugs go through four nymph stages.
In each new stage, they look more like an adult.
How many different nymph stages can you find?

Thinking about Insect Life Cycles

1. Tell about the life cycle of a bee.

2. Tell about the life cycle of a milkweed bug.

Life Goes Around

On a lucky day, you might see a ladybug.
A ladybug is red with black spots.
This is an adult ladybug.
But have you ever seen a baby ladybug?

Adult ladybugs mate.

Then, the female lays eggs.

When an egg hatches, a larva comes out.

The black larva is a baby ladybug.

But it doesn't look like its parents.

The larva eats and grows for about 4 weeks.

Then, the larva **pupates**.

Inside the pupa, the larva is changing.

When the pupa opens, an adult ladybug comes out.

It is red with black spots.

Now it looks just like its parents.

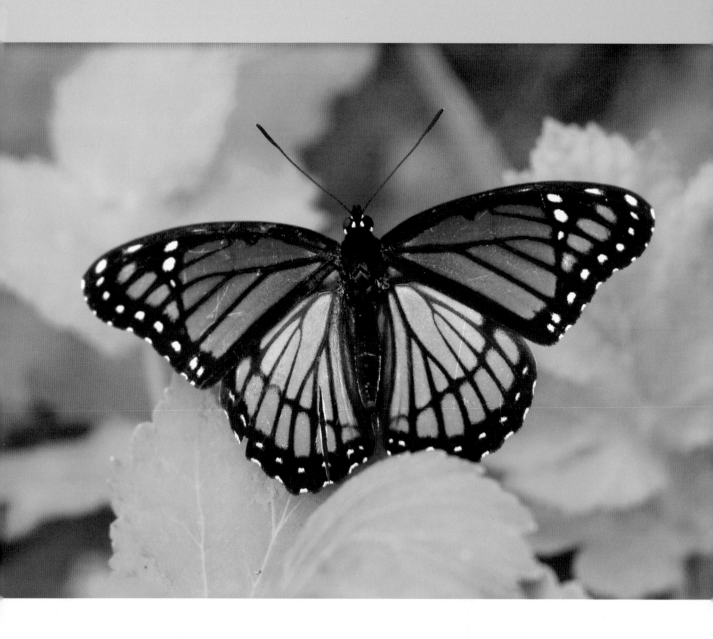

The ladybug life cycle is like the life cycle of many other insects.

It is like the life cycle of mealworms.

It is like the life cycle of butterflies and moths.

But it is different from the life cycle of some other animals.

Some animals hatch from eggs.
Some animals are born alive.
They all grow up to be adults.
The adults mate and have babies called **offspring**.

Every animal goes around the life cycle.
Cycle means to go around.
The life cycle of a robin looks like this.

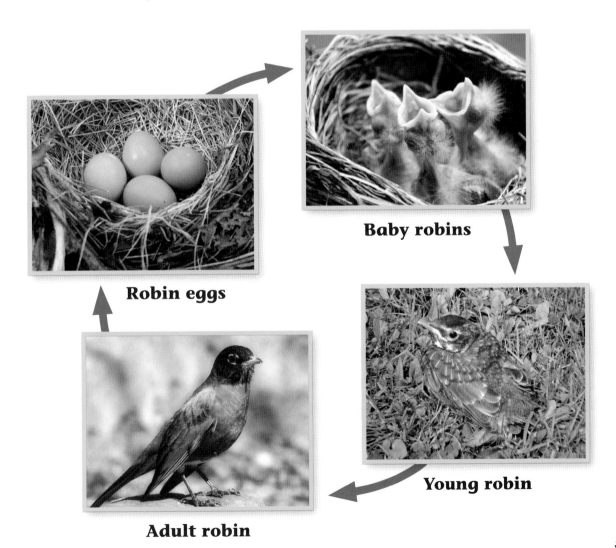

Robin eggs

Baby robins

Young robin

Adult robin

Trout lay eggs in streams.
After 6 to 8 weeks, the eggs hatch.
Tiny, fat babies swim out.
You can see that they are fish.
But they don't look like their parents yet.

For the next year, they grow up little by little.
In 2 years, they are adults.
They look just like their parents.
They mate and lay eggs in streams.
Can you describe the trout life cycle?

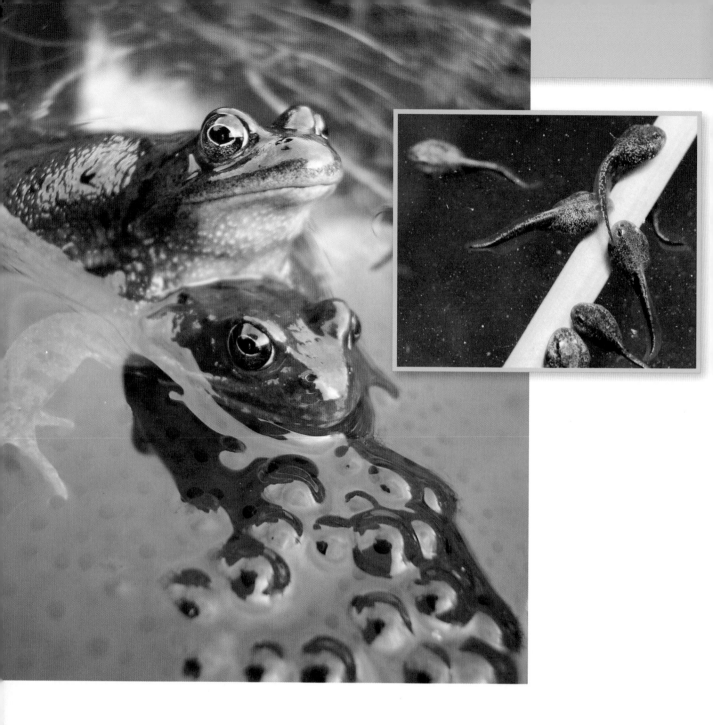

Frogs lay eggs in water, too.
When an egg hatches, a tadpole swims out.
It looks more like a fish with a big head
than a frog.
It doesn't look like its parents yet.

The tadpole eats and grows.
In a few weeks, the tadpole starts to change.
Its long, flat tail gets shorter.
Its legs start to grow.

In a few more weeks, the tadpole has grown into a frog.
It looks just like its parents.
Can you describe the frog life cycle?

Ducks lay eggs in a nest in a **marsh**.

The mother duck sits on the eggs to keep them warm.

When they hatch, the babies are fluffy and yellow.

The babies are called ducklings.

You can see that they are ducks.

But they don't look like their parents yet.

The ducklings eat and grow.

In a few weeks, they get their brown feathers.

In a few months, they are adults.

They look just like their parents.

In the next year, the adult ducks will mate.

They will raise new families of ducklings.

Can you describe the duck life cycle?

Mice do not lay eggs.

Baby mice grow inside the mother.

The babies are born alive.

Newborn mice are pink, hairless, and blind.

You can see that they are mice.

But they don't look like their parents yet.

In a few days, the babies open their eyes.
Their fur starts to grow.
In a few months, they will be adults.
They will be ready to continue the life cycle.
They will have babies of their own.
Can you describe the life cycle of mice?

Thinking about Life Goes Around

1. Does a ladybug larva look like its parents?

2. Tell about the life cycle of a ladybug.

3. Tell about the life cycle of a different animal.

4. Name five animals that hatch from eggs.

5. Name three animals that are born alive (not from eggs).

Fossils

How do we know what plants and animals
looked like millions of years ago?
We look at **fossils**.
Fossils are the remains of plants and animals
that lived a long time ago.
Scientists study fossils to learn about the past.

Dinosaurs lived a long, long time ago.
No dinosaurs are living today.
But scientists can study dinosaur fossils to learn
about them.

One of the most famous dinosaur fossils
is named Sue.
Sue is a *Tyrannosaurus rex*.
This is what Sue might have looked like
when she was found in South Dakota.

Scientists dug Sue out of the ground very carefully.

In the lab, Sue's fossil bones were carefully cleaned.

After a lot of hard work, all of Sue's bones and teeth were ready to put together.

It took a long time to get all the fossil bones
in the right places.
Finally, the bones all fit together.
Everyone can now see what Sue's skeleton
looked like.

After the skeleton was together, scientists wanted to know what Sue looked like when she was alive.

They used pretend muscles, skin, and eyes on a copy of her skeleton to make her look real.

This is what Sue might have looked like.

Dinosaurs have not lived on Earth
for millions of years.
But animals that look like dinosaurs
are found on Earth today.
Lizards, such as iguanas, look like
dinosaurs.

Insects also lived when dinosaurs
roamed Earth.
Fossils of insects are found in rock.
Insect fossils are found in **amber**, too.

Amber is very old **resin** from a tree.
Resin is a very sticky liquid that comes from some kinds of plants.
If you have ever touched pine tree sap, then you know how sticky resin can be.

If an insect landed in resin, it was trapped
and died.
Millions of years later, those insects are
fossils in amber.
Bees, dragonflies, and wasps are some of
the insect fossils found in amber.

Thinking about Fossils

1. What are fossils?

2. Where are fossils found?

3. What do fossils tell us?

4. Why are fossils important?

Glossary

adult a fully grown organism **(16)**

amber a hard fossil resin from a tree **(75)**

basic need something that is needed for plants and animals to survive. Plants and animals need air, water, food, space, shelter, and light. **(4)**

egg the first stage of a life cycle **(15)**

flower the part of a plant that grows into fruit **(18)**

fossil a part of a plant or animal that lived long ago. Fossils can be bones, shells, or leaves. Fossils can also be tracks or burrows of past life on Earth. **(67)**

fruit the part of a plant with seeds in it. Flowers grow into fruit, and fruit produce seeds in plant reproduction. **(22)**

habitat the place or natural area where plants or animals live **(3)**

insect an animal that has six legs and three main body parts. They are the head, thorax, and abdomen. **(5)**

larva (plural **larvae**) a stage in the insect life cycle after hatching from eggs. Insect larvae look different from their parents and are often wormlike. **(16)**

life cycle the stages in the life of a plant or animal **(18)**

living alive (3)

marsh soft, wet land that is sometimes covered with water (62)

nutrient something that living things need to grow and stay healthy (6)

nymph a stage in the insect life cycle that has no larva or pupa. Nymphs look like their parents, but are smaller. (50)

offspring a new plant or animal produced by a parent (57)

organism a living thing. Plants and animals are organisms. (32)

pupa (plural **pupae**) a stage in the insect life cycle between the larva and adult stages (44)

pupate to change into a pupa (55)

reproduction the process of producing offspring (22)

resin a sticky liquid that comes from some plants (76)

seed the part of a plant found inside fruit. Seeds can grow into new plants. (10)

survive to stay alive (11)

thrive to grow fast and stay healthy (17)

variation difference (32)